HARVESTING FOG
FOR WATER

BY CECILIA PINTO McCARTHY

CONTENT CONSULTANT
Hosein Foroutan, PhD
Assistant Professor, Department of Civil and Environmental
Engineering, Virginia Tech University

Core Library

Cover image: People need training to set up and
maintain fog nets.

An Imprint of Abdo Publishing
abdobooks.com

abdocorelibrary.com

Published by Abdo Publishing, a division of ABDO, PO Box 398166, Minneapolis, Minnesota 55439. Copyright © 2020 by Abdo Consulting Group, Inc. International copyrights reserved in all countries. No part of this book may be reproduced in any form without written permission from the publisher. Core Library™ is a trademark and logo of Abdo Publishing.

Printed in the United States of America, North Mankato, Minnesota
042019
092019

Cover Photo: Desiree Martin/AFP/Getty Images
Interior Photos: Desiree Martin/AFP/Getty Images, 1, 25; Mariana Bazo/Reuters/Newscom, 4–5, 6, 9, 30; Christina Prinn/iStockphoto, 12–13; Probal Rashid/LightRocket/Getty Images, 14; iStockphoto, 17 (land), 22–23, 29 (nets), 45; Rostik Solonenko/Shutterstock Images, 17 (clouds); Shutterstock Images, 18, 20; Martin Bernetti/AFP/Getty Images, 27, 32–33, 43; Red Line Editorial, 29 (droplets); Georg Ismar/picture-alliance/dpa/AP Images, 34–35

Editor: Marie Pearson
Series Designer: Ryan Gale

Library of Congress Control Number: 2018966160

Publisher's Cataloging-in-Publication Data

Names: McCarthy, Cecilia Pinto, author.
Title: Harvesting fog for water / by Cecilia Pinto McCarthy
Description: Minneapolis, Minnesota : Abdo Publishing, 2020 | Series: Unconventional science | Includes online resources and index.
Identifiers: ISBN 9781532119002 (lib. bdg.) | ISBN 9781532173189 (ebook) | ISBN 9781644940914 (pbk.)
Subjects: LCSH: Water harvesting--Juvenile literature. | Precipitation trapping--Juvenile literature. | Water-supply--Management--Juvenile literature. | Environmental protection--Juvenile literature. | Water cycle--Juvenile literature. | Water conservation--Juvenile literature.
Classification: DDC 628.1--dc23

CONTENTS

CHAPTER ONE

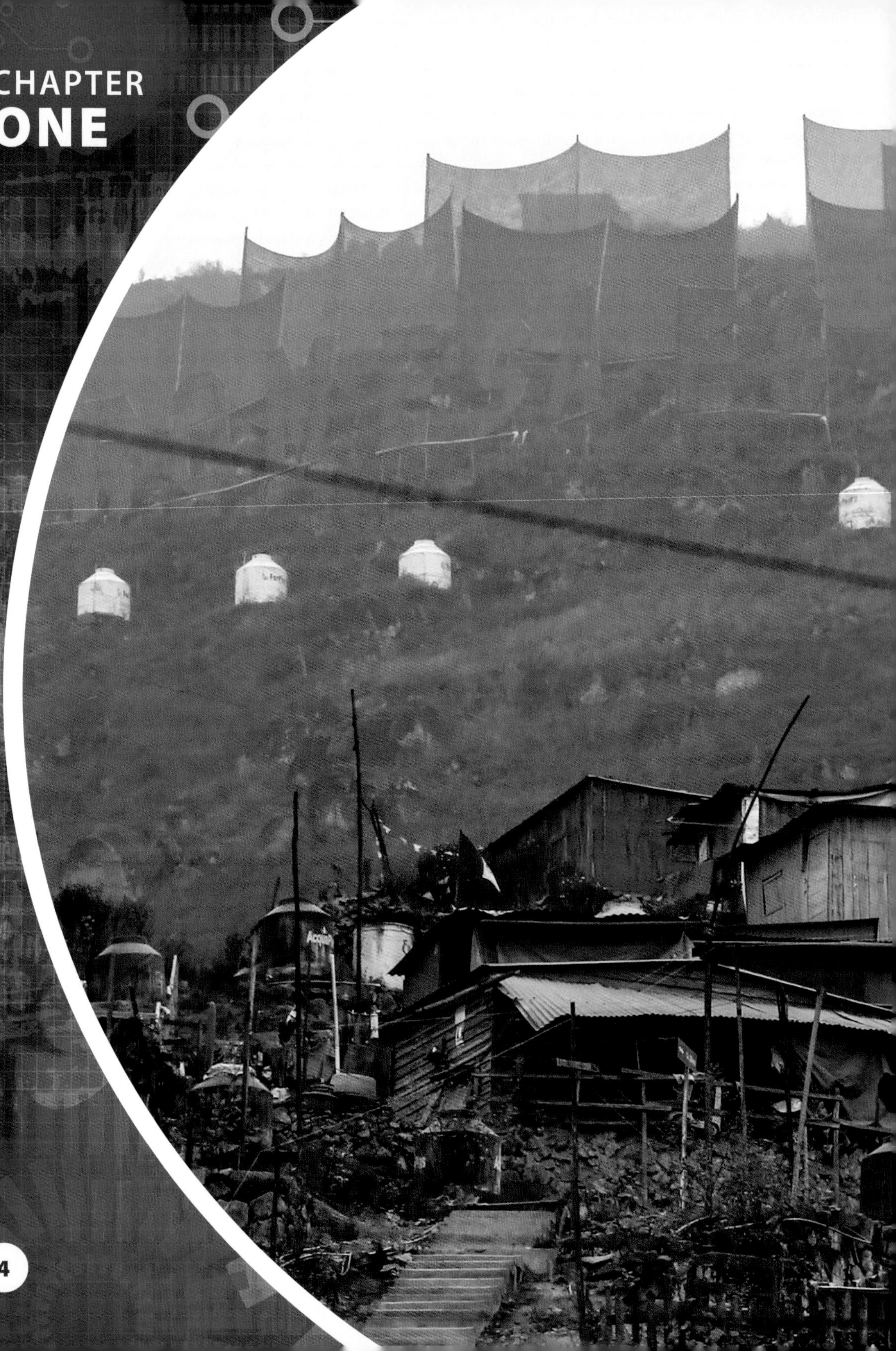

WATER FROM CLOUDS

On the outskirts of Lima, Peru, Abel Cruz Gutierrez hikes to the top of a steep hill. The land at his feet is dry and rocky. Gutierrez is barely visible through a blanket of fog. All around him, panels of netting stretch out between poles. Strong cables anchor the poles in place. Gutierrez stops at a panel and tugs at the cables. He checks that the poles are properly secured to the ground. He inspects the netting for tears. As Gutierrez runs his hands along the nylon mesh, they become wet with water.

Fog nets are an important source of water in many communities.

Without water from the fog nets, growing food would be difficult and expensive where Avalos lives.

A network of pipes leads away from the panels to a white storage tank. Water trickles from two pipes into the tank. Downhill, Gutierrez stops to speak with María Teresa Avalos. She is soaking a garden with water from the storage tank. The ground on this part of the hillside is lush with vegetation. Parsley, cilantro, potatoes, and celery grow in her garden. Avalos harvests armfuls of vegetables. She will use the foods to make soup for dinner.

WATER IN A DESERT

Lima is one of many locations around the world where water is scarce. The capital city is in a desert on the western coast of Peru. Here it almost never rains. Living in such a dry area is difficult. It's especially hard for people in poverty like Avalos. She is one of approximately 3 million people who live on the outskirts of Lima. These communities have no running water. Trucks bring in water. People must buy the water they need to drink and do chores. The water is expensive. They cannot afford much. The water is not always safe to drink.

Gutierrez helps bring water to Peruvians who lack running water. He leads an organization called *Movimiento Peruanos sin Agua* (Peruvians without Water Movement). The group places fog nets in hillside villages not

ECONOMIC BENEFITS

In some communities, people use fog water to make products to sell. In Majada Blanca, Chile, fog water irrigates olive trees. The olives are processed and sold as organic olive oil. The first crop was harvested in May 2017. Residents used some of the profit to install more nets and plant more trees.

far from the coast. In the hills, the humid air from the Pacific Ocean cools down. This creates a thick blanket of fog. When the fog touches the nets, it condenses into water droplets. The droplets flow down the net. They pool together in a trough that runs along the bottom of the net. The water then flows downhill into pipes. The pipes empty into storage tanks on the hillside.

Usually, water from fog goes unused by people. Fog nets turn fog into a free source of water. With fog water, people can water crops. These crops feed their families and provide a source of income. They use the water for drinking, bathing, and household cleaning. Their farm animals have water to drink. Fog does not carry many contaminants. Dust also does not tend to stick to the fog nets. This means harvested water is drinking quality.

SIMPLE AND SUSTAINABLE

People harvest water from fog in many dry climates around the globe. There are nets in Chile, Morocco,

Water collected from fog nets can be stored in barrels for later use.

Ghana, Eritrea, and South Africa. In the United States, fog is harvested on the coast of California. Fog water benefits the environment as well as people. It is used to grow trees and restore native plants.

Harvesting water from fog has many advantages. Fog collecting is sustainable. That means the water does not run out. Also, collecting water from fog does not harm the environment. Collecting fog usually does not require energy. Setting up fog nets is inexpensive. The nets and other parts of harvesting systems such as poles, pipes, and storage tanks are fairly cheap. Once a fog harvesting system is in place, it is easily maintained. Area residents are trained to inspect and care for the nets and other equipment.

Harvesting water from fog, however, does have some issues. Wind, temperature, and altitude influence the water content

COPYING NATURE

Many desert plants and animals survive by collecting fog water. On foggy mornings, a beetle in Africa's Namib Desert climbs to the top of a dune. It lifts the back end of its bumpy body into the fog. The tops of the bumps are glassy. They attract water droplets. The droplets grow large and roll into the waxy area between the bumps. Then they glide down into the beetle's mouth. The desert Tortula moss uses tiny fibers on its leaves to capture fog water. Researchers study these animals and plants to improve fog collecting technology.

of fog. Nets must be set up in specific locations where there is consistent fog. Equipment can be damaged by strong winds. Soil particles may clog spaces in the mesh.

Researchers are creating more efficient and effective fog collectors. Improved fog collector designs gather more water. Systems must be durable to withstand harsh environments. Stronger and more efficient fog collection systems will provide more water to dry areas worldwide.

EXPLORE ONLINE

Chapter One discusses how fog nets provide water for people in dry areas around the world. Go to the website below to read how fog collectors changed people's lives in the village of Chungungo, Chile. What new information did you learn from the website? How is the information from the website like the information in Chapter One?

CASE STUDIES: HARVESTING WATER FROM THE SKY
abdocorelibrary.com/harvesting-fog

A SOLUTION TO WATER SCARCITY

Nearly three-quarters of Earth is covered by water. This seems like a lot of water. But most of it is salt water. Only 3 percent is fresh water. Most fresh water is frozen as glaciers and polar ice. That leaves just 1 percent available for people to use. Fresh water is replenished by precipitation such as rain and snow. Many places have water scarcity. In these areas, the supply of water cannot meet the needs of the people. Experts acknowledge that water scarcity is a worldwide problem. And it's getting worse. One study

Most of the fresh water readily available to humans is found in rivers and lakes.

Many people around the world don't have access to clean water.

estimates that half a billion people worldwide cope with year-round water scarcity.

WHAT CAUSES WATER SHORTAGES?

Many things cause water shortages. Human population growth stresses water supplies. People may overuse freshwater sources such as rivers and lakes. These sources dry up. Sometimes humans pollute water sources. Then water becomes unusable. Some people live in areas that naturally have less water. Deserts, for example, get little or no precipitation.

Climate change is also reducing global water supplies. Since the late 1800s, Earth's average surface temperature has risen. This and other factors cause extreme weather events. Some areas experience floods. Heavy rainfall can pollute water supplies. It can damage pipes and storage tanks. In other areas, droughts cause lengthy water shortages.

Water shortages cause health crises. Without clean water for bathing and drinking, people get sick. In some places, people must walk many miles

DISAPPEARING FOG

Researchers at the University of California, Berkeley, discovered that the number of foggy days in California's Central Valley has been declining. They tracked the number of foggy days over 32 winters. The number of days had dropped by 46 percent. An increase in droughts has resulted in less fog. Fewer foggy days are harming the growth of the region's fruit and nut trees. The trees need a cold winter period to rest. Without the usual thick fog, the number of chilly winter days has decreased. Trees that do not rest enough yield less fruit.

to collect water. Usually, women and girls are burdened with the daily task of fetching water for their families. This means they miss out on other opportunities. Children miss classes or cannot attend school at all. When stored, water may become contaminated. Unsafe water causes illness and death. Harvesting fog for water can lessen the problems associated with water scarcity. Harvesting fog brings water to communities where the resource is scarce.

THE WATER CYCLE

Water exists in different forms, or states. It may be a liquid, solid, or gas. Liquid water is found in lakes and rivers and as rain. Snow and ice are solid forms of water. Water as a gas is called water vapor. Water moves on, above, and below Earth's surface. This continuous movement of water is called the water cycle. As water moves, it changes states. Heat from the sun causes liquid water to evaporate and become a gas. The water vapor rises into the atmosphere. There it can travel thousands of miles. Then the vapor cools and

THE WATER
CYCLE

The diagram below shows how water moves around Earth. Can you identify when water changes between liquid and gaseous states as it cycles?

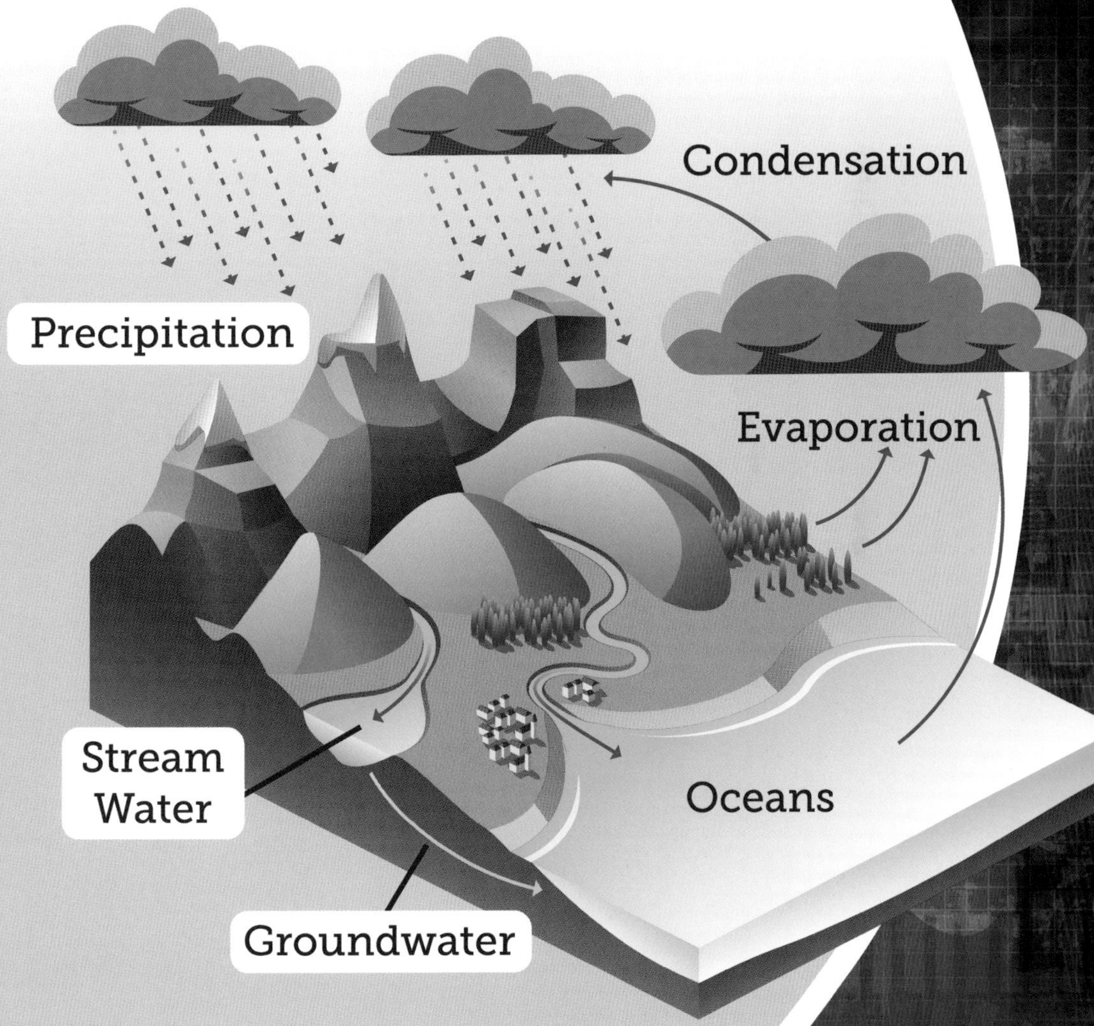

Condensation

Precipitation

Evaporation

Stream Water

Oceans

Groundwater

condenses. The liquid droplets combine with microscopic particles in the air. As the water droplets combine and grow larger, they form clouds.

Fog is a type of cloud that forms near the ground instead of in the sky. Fog occurs when the air becomes saturated. Saturated air can no longer hold any more water vapor. Cool air near Earth's surface causes the water vapor to condense.

DEFORESTATION CAUSES DROUGHTS

Trees and other plants play an important role in the water cycle. Trees take in water through their roots. They release water vapor into the atmosphere through pores on their leaves. One tree can give off thousands of gallons of water in one year. The water vapor forms clouds that then create rain. The process also helps cool the air. In many areas worldwide, people are clearing forests. The land is farmed or paved into roads. With fewer trees adding water to the air, there is less rain. This causes droughts.

Water vapor in the air condenses into fog in the same way it condenses on a cold glass of soda on a hot, humid day.

In deserts, there can be times when fog is the only source of water.

Fog forms when warm, humid air passes over cold ocean water or land. Coastal areas become foggy when warm air travels over cold ocean waters. Wind currents blow the fog inland. In deserts, heat from the sun warms the air during daytime hours. When the temperature drops at night, the warm air vapor condenses to form fog.

STRAIGHT TO THE
SOURCE

A 2018 *National Geographic* article described the global water crisis:

> *Fourteen of the world's 20 megacities are now experiencing water scarcity or drought conditions. As many as four billion people already live in regions that experience severe water stress for at least one month of the year. . . . Nearly half of those people live in India and China. . . .*
>
> *Disaster data compiled by the U.N. clearly shows floods are also getting worse. They are happening more frequently, especially in coastal regions and river valleys, and affecting more people. . . .*
>
> *Humanity is facing a growing challenge of too much water in some places and not enough water in others. This is being driven not just by climate change, but by population and economic growth and poor water management.*

Source: Stephen Leahy. "From Not Enough to Too Much, the World's Water Crisis Explained," *National Geographic*. National Geographic, March 22, 2018. Web. Accessed October 3, 2018.

What's the Big Idea?
Review the passage closely. What are some of the water challenges people are facing? What is causing the problems?

CHAPTER THREE

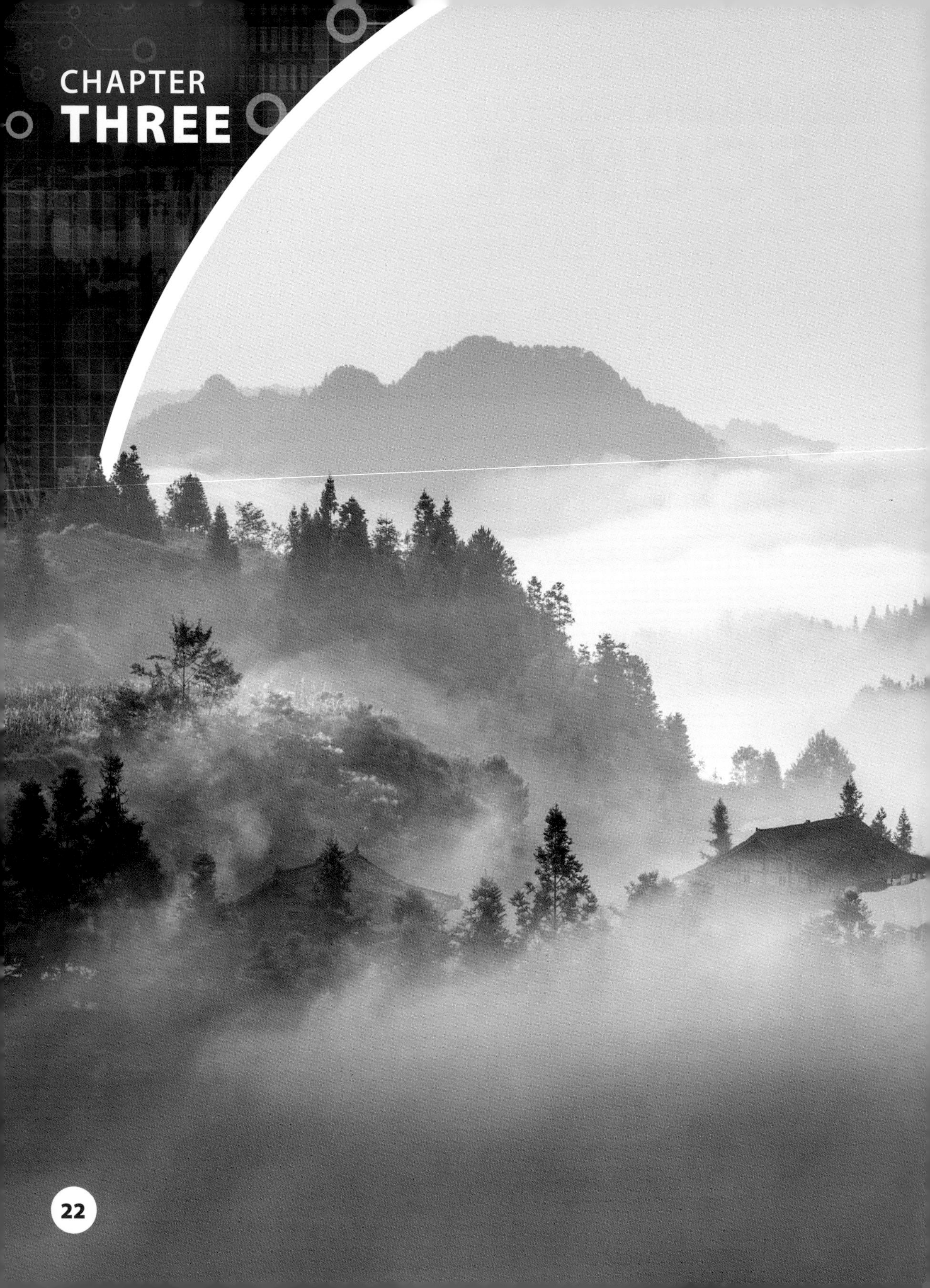

FOG COLLECTORS

Fog harvesting systems only work in certain places. Several factors determine whether an area will be a good fog collection site. A site needs frequent fog. Dry and semidry tropical and subtropical climates tend to get more fog. It forms in these areas when warm, humid air passes over cool land and ocean surfaces. Fog also forms in coastal areas with hills, dunes, or mountains. The cold temperature at high altitudes causes warm air to condense. The fog is thick.

Wind patterns play an important role. Steady winds that come from one direction are

In order for a fog net to be useful, a place needs to get fog often.

best for fog collection. Winds blowing in from the ocean are ideal. Collection nets are installed facing into the wind. That way, the most fog hits the nets as it blows in. High, steady winds ensure that fog will hit the collection nets. Nets set up at high altitudes must be strong. Otherwise they can be damaged by strong winds. Chile, Peru, and Morocco are coastal countries that successfully harvest fog for water.

FOGQUEST

FogQuest is a nonprofit organization based in Canada. It started in 2000. The organization works with companies, government agencies, and other groups to install fog collectors around the world. FogQuest has collectors in Chile, Guatemala, Ethiopia, Nepal, Eritrea, and Morocco. Each site is evaluated to make sure that there is enough fog to harvest. Then FogQuest works with people in the local community to install a collection system. It also trains residents to operate and maintain the system.

FOG COLLECTION SYSTEMS

Harvesting fog for water is simple and inexpensive. It does not need external

Wind helps blow fog into nets.

power. Wind and gravity do all the work. Wind blows fog into the nets. Gravity pulls water from nets into pipes, storage containers, and nearby communities. Harvesting systems rely on just a few pieces of basic equipment. It costs approximately $75 to $200 to build a small fog collector. One large collector costs between $1,000 and $1,500.

A standard fog collector (SFC) is installed to evaluate a site before a large system is built. Researchers test how much water can be collected at the location. They learn which times of year are most productive. Ideally, the evaluation process takes at least one year. If enough water is collected with an SFC, a larger fog collection system will be erected.

An SFC is made of a ten-square-foot (1 sq m) metal frame. A double layer of plastic mesh called raschel mesh stretches tightly across the frame. The durable flat mesh fibers and plastic surface easily capture fog. Water drains from the net when the two layers of mesh rub against each other. Spaces in the mesh allow wind to pass through. Without spaces, the panel would form a wall against the wind. The wind would blow the fog around the panel. The bottom of the frame is 6.6 feet (2 m) above the ground. Posts on either end of the frame anchor it to the ground. Underground, the

SFCs help people know if fog nets will be useful in an area before they spend a lot of money on larger nets.

HARVESTING FOG AT SCHOOL

In 2014, fog collectors were installed at the Ilmasin Primary School in Kenya. The school is near the Ngong hills. The area receives little or no rain for six months of the year. But there is plenty of fog in the morning and at night. The nets provide safe, reliable water for the students. At first, the nets produced approximately 16 gallons (60 L) of water a day. Then in 2015, strong winds destroyed part of the system. Daily water dropped down to 5 to 8 gallons (20–30 L) per day.

posts are cemented in place. Additional cables help support the frame.

A collection trough runs along the bottom of the frame. The trough is slightly longer than the frame on both ends. This design ensures that it catches the maximum amount of water dripping from the mesh. The trough also catches rain or water that may blow off the frame. One end of the trough slopes downward. Gravity drains the water toward a plastic tube on the lower end of the trough. From there, the water travels through a pipe to a storage container.

HOW NETS
TRAP FOG

The diagram below shows how fog particles travel through netting. Why is it important for the net to both attract and shed water? What would happen to water droplets if the openings were too small?

FOG PARTICLE

Water forms around a particle like sand.

FOG COLLECTION

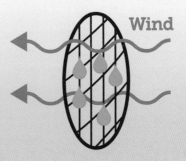

Wind

Wind carries fog to the net. The net captures the fog particles.

FOG PARTICLE PATHS

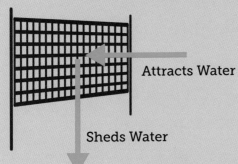

Attracts Water

Sheds Water

Net needs to both attract and shed water.

If the net doesn't allow enough wind through, the wind will go around the net, carrying the fog with it.

Several LFCs can provide water for small communities.

LARGE FOG COLLECTORS

Large fog collectors (LFCs) are set up in areas where SFCs have successfully harvested plenty of water. LFCs are larger than SFCs. They are approximately 13 feet (4 m) high and 33 feet (10 m) or 39 feet (12 m) wide. LFC mesh is not attached to a rigid frame. Instead, the mesh panel is secured to the poles with cables. One LFC can collect up to 130 gallons (500 L) of water in a single foggy day. On average, an LFC yields 50 gallons (200 L) of water per day.

Fog collection is most successful in regions where local people are actively involved in the project at every stage. They help plan, install, operate, and maintain the collection system. Maintenance of a collection system is simple. Nets, cables, and other parts of the system are checked frequently for rips, fraying, and other issues. Debris such as plant matter and soil is cleared from collection troughs and pipes. Storage tanks must be kept free of algae and other things that may contaminate the water.

FURTHER EVIDENCE

Chapter Three discusses the equipment and environmental conditions needed to collect water from fog. Review the chapter and identify one of its main points. What key evidence supports this point? Go to the website below and explore more ideas about harvesting water from fog. Find a quote from the website that supports this point.

ADAPTATION STORIES FROM AROUND THE GLOBE
abdocorelibrary.com/harvesting-fog

CHAPTER FOUR

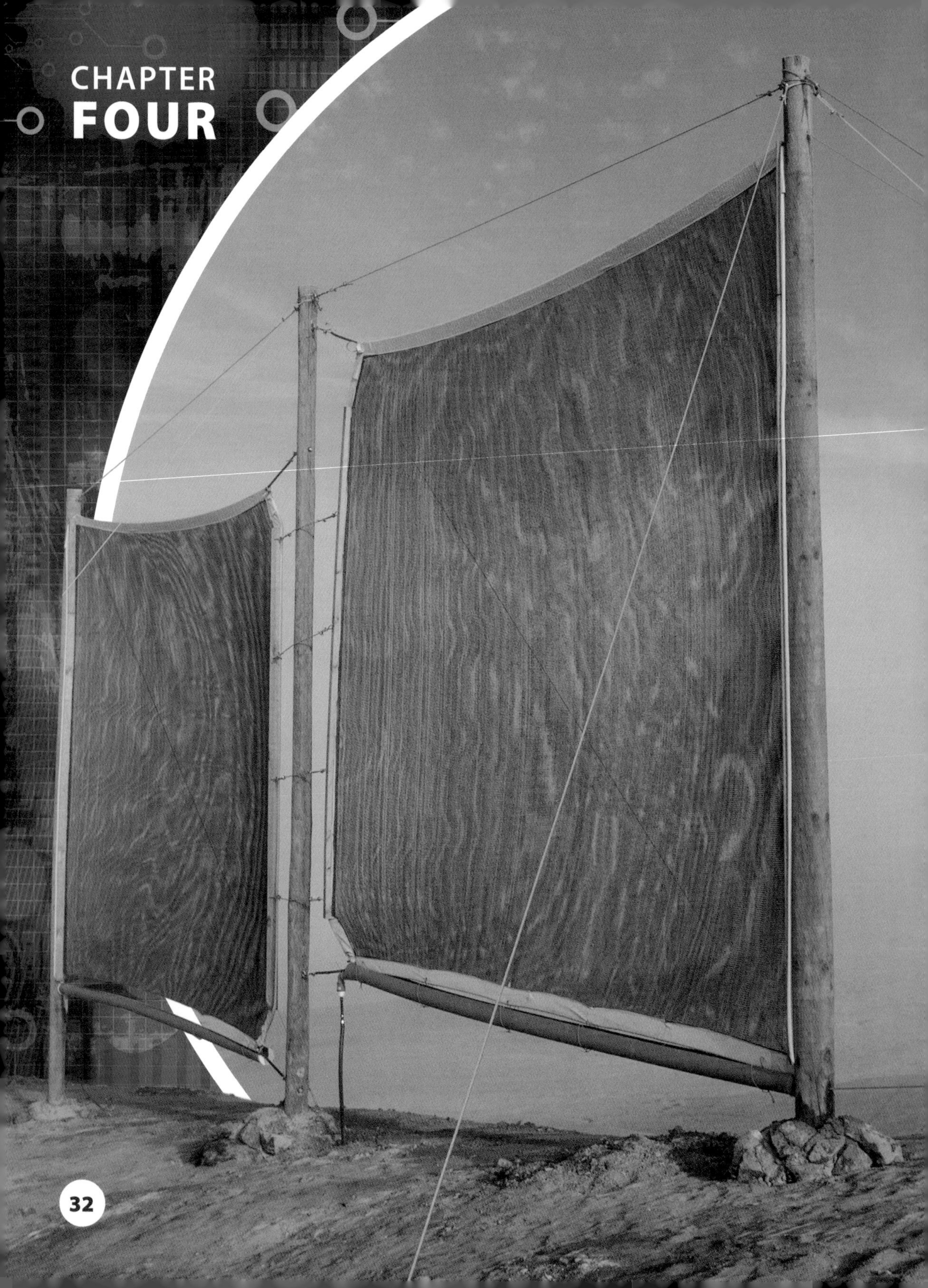

THE FUTURE OF HARVESTING FOG

People have collected water from fog for at least 1,000 years. But the first large-scale project began in 1985 in South America's Atacama Desert. The Atacama is one of the driest places on Earth. Some parts of the desert have never had a drop of rain. But the Atacama gets fog. In Chile, a low, dense fog called camanchaca moves inland from the cold waters of the Pacific Ocean. The fog passes over the mountain El Tofo. In 1987, a test project called the Camanchaca Project began. It placed 50 fog collectors on El Tofo. Researchers tested designs for an inexpensive

Fog nets are used in desert areas in Chile.

Many fog nets can be used to provide water for larger communities.

and easy-to-operate fog collector. The project lasted until 1991. The collectors successfully produced several gallons of water each day. The water was used at a forestry plantation.

In 1992, several organizations provided funds to build a buried pipeline. The pipeline connected the collection site to the coastal fishing village of Chungungo. Before the pipeline, water was trucked

into the village. Villagers stored the water in old oil drums.

Over time, more fog collectors were built on El Tofo. The collectors provided villagers with an average of 4,000 gallons (15,000 L) of water each day. The project lasted for ten years. It became a model for the development of fog collection systems around the world.

THE LARGEST SYSTEM

In 1989, Aissa Derhem, President of Morocco's Dar Si Hmad Foundation (DSH), learned about the Camanchaca Project. He wanted to harvest fog water for people in drought-stricken areas of southwest Morocco. In 2006, he installed the first SFC on Mount Boutmezguida. The nearby area borders the Sahara Desert. For centuries, residents relied on rainwater or well water. But frequent droughts and overuse meant wells were often dry. With 143 foggy days per year, the region is ideal for fog collection.

The Moroccan project became the world's largest fog harvesting system. Fifteen fog collection units harvest fog that rolls in from the Atlantic Ocean. On average, 3,170 gallons (12,000 L) of water are collected each day. The water is stored in reservoirs and piped into area homes. People pay a small fee for the water they use.

A BETTER DESIGN

Researchers are seeking ways to improve fog collection systems. Worldwide, scientists at companies, universities, and other organizations are developing new ways to harvest fog. Some focus on finding better net materials and mesh weave patterns. Others experiment with refining the system designs. They hope to create more efficient and effective fog harvesting systems.

The fog harvesting systems currently on Mount Boutmezguida are called CloudFishers. The nets have rubber parts that make them flexible. They can

CHANGING LIVES

When CloudFisher nets came to Morocco, women and girls no longer spent hours fetching water. They could use their free time on education and other activities. DSH trained women to manage the fog collectors. They use mobile phones to relay information about the system. DSH holds classes, including reading and writing. It also runs a Water School. The school uses science, technology, art, and math to educate children about water.

withstand winds of up to 75 miles per hour (120 km/h). CloudFisher nets have woven mesh that traps a lot of fog. The net's grid pattern stops it from bulging in the wind and draining water outside the trough.

More CloudFishers were installed on the mountain during 2018.

In 2018, scientists at Virginia Polytechnic Institute and State University presented a new fog collector design called the Harp. Unlike traditional fog nets, the Harp uses vertical wires instead of mesh. When fog condenses on mesh, water droplets can get stuck in the

SLIPPERY GROOVES

As wind blows fog into a fog net, the mesh must hold onto the water droplets. It must also be able to shed the droplets so that they fall into the trough. It's tricky to find a material that will both hold and shed water droplets. In 2018, scientists developed a surface that attracts and sheds water more efficiently than mesh. The surface has parallel microscopic grooves. A special oil on the grooves attracts water droplets. The grooves and coating also help the water trickle down the surface.

holes. On the Harp, droplets can freely drip down the vertical wires without stopping. During one test, a Harp collected three times more water than a fog net.

Scientists at the Massachusetts Institute of Technology are using electrical forces to pull water droplets toward a fog collector's mesh. They place a device called an ion emitter near a fog collector. The emitter makes an electrical field that charges water droplets in the foggy air. This causes the droplets to be attracted to wire mesh. Even droplets that pass through the mesh get pulled back and trapped. The voltage can be increased or decreased according to wind speed and other conditions.

A SUSTAINABLE SOLUTION

The process of harvesting fog has remained virtually the same for decades. It provides a sustainable solution to water scarcity. Much of the success of fog collection lies in its simplicity.

But fog collection has some limitations. Enough water can only be collected in specific areas and under certain conditions. Site testing may take months or years to complete. After nets are installed, the amount of fog water collected may be inconsistent. High winds and stormy weather take a toll on nets and other parts of the system. Fog water may be blown away from the net. Spaces in the mesh may clog with droplets and slow collection. People must work together to care for the system.

As the world's population booms and Earth's climate changes, the demand for water grows. Harvesting fog for water is an eco-friendly and cost-effective solution. New methods and technologies will ensure that people continue to harvest water from the clouds.

STRAIGHT TO THE
SOURCE

Researchers have created a water harvester that uses a metallic powder that acts like a sponge. Journalist Devin Coldewey described it:

> It's essentially a powder made of tiny crystals in which water molecules get caught as the temperature decreases. Then, when the temperature increases again, the water is released into the air again. . . .
>
> They put together a box about two feet [0.6 m] per side with a layer of MOF [metal-organic framework] on top that sits exposed to the air. Every night the temperature drops and the humidity rises, and water is trapped inside the MOF; in the morning, the sun's heat drives the water from the powder, and it condenses on the box's sides, kept cool by a sort of hat. The result of a night's work: 3 ounces [85 g] of water per pound [0.5 kg] of MOF used.
>
> Source: Devin Coldewey. "This Box Sucks Pure Water out of Dry Desert Air." *TechCrunch*. Oath Tech Network, June 2018. Web. Accessed October 3, 2018.

Back It Up

Based on this passage and what you have learned in this book, how is this water harvester similar to a fog collector net? In what ways is it different?

FAST FACTS

- Just 1 percent of all the water on Earth is available to be used by people.

- Water scarcity happens when people's need for water exceeds the supply.

- Fog is a type of cloud that forms near the ground.

- Some plants and animals that live in dry regions are adapted to collect water from fog.

- Climate conditions and geography determine the best sites for fog collectors.

- Wind pushes fog against fog nets, where it condenses into droplets.

- Gravity moves water along the fog collector through pipes and into storage areas.

- The Camanchaca Project in Chile was the first large-scale attempt to harvest fog for water.

- The largest fog harvesting system is in Morocco.

- In some communities, fog harvesting frees women and girls from traveling for hours to retrieve water.

- New research focuses on improving fog collection systems so that they are more efficient and collect as much water as possible.

STOP AND
THINK

Tell the Tale

Chapter Three describes fog collection systems and the ideal sites where they can be set up. Imagine you are touring a fog collection system on a mountaintop. Write 200 words describing the collection process. What do you see? What steps are used to harvest fog and collect water? Use several details to describe what is going on at the site.

Surprise Me

Chapter Two talks about water scarcity. After reading this book, what two or three facts about fog and water did you find most surprising? Write a few sentences about each fact. Why did you find each fact surprising?

Dig Deeper

After reading this book, what questions do you still have about harvesting fog for water? With an adult's help, find a few reliable sources that can help you answer your questions. Write a paragraph about what you learned.

Say What?

Studying how water is harvested from fog can mean learning a lot of new vocabulary. Find five words in this book you've never heard before. Use a dictionary to find out what they mean. Then write the meanings in your own words, and use each word in a new sentence.

GLOSSARY

climate
the average, slowly varying weather conditions in a certain region

climate change
the changes in Earth's climate because of chemical changes in the atmosphere that have led to rising temperatures

condense
to change states from a gas to a liquid

evaporate
to change from a liquid into a gas

harvest
to gather a product

humid
having a great deal of moisture

irrigate
to use human-made methods to water plants or crops

plantation
an area where trees grow under human supervision

poverty
the state of having little money and few possessions

saturated
holding as much water as possible

sustainable
able to be maintained at a certain rate or level

ONLINE RESOURCES

To learn more about harvesting fog for water, visit our free resource websites below.

Visit **abdocorelibrary.com** or scan this QR code for free Common Core resources for teachers and students, including vetted activities, multimedia, and booklinks, for deeper subject comprehension.

Visit **abdobooklinks.com** or scan this QR code for free additional online weblinks for further learning. These links are routinely monitored and updated to provide the most current information available.

LEARN MORE

McCarthy, Cecilia Pinto. *Supplying Water for a City.* Minneapolis, MN: Abdo Publishing, 2019. Print.

Mulder, Michelle. *Every Last Drop.* Victoria, BC, Canada: Orca, 2014. Print.

INDEX

About the Author

Cecilia Pinto McCarthy has written several children's books about science and nature. She also teaches classes at a nature sanctuary. She and her family live north of Boston, Massachusetts.